Editor
Lorin Klistoff, M.A.

Managing Editor
Karen Goldfluss, M.S. Ed.

Illustrator
Blanca Apodaca

Cover Artist
Brenda DiAntonis

Art Manager
Kevin Barnes

Art Director
CJae Froshay

Imaging
Alfred Lau

Publisher
Mary D. Smith, M.S. Ed.

Author

Bev Dunbar

(Revised and rewritten by Teacher Created Resources, Inc.)

This edition published by *Teacher Created Resources, Inc.*
6421 Industry Way
Westminster, CA 92683
www.teachercreated.com

ISBN-1-4206-3528-X

©2005 Teacher Created Resources, Inc.

Made in U.S.A.

with permission by Blake Education

The classroom teacher may reproduce copies of materials in this book for classroom use only. The reproduction of any part for an entire school or school system is strictly prohibited. No part of this publication may be transmitted, stored, or recorded in any form without written permission from the publisher.

Table of Contents

Introduction 3
How to Use This Book 4
Exploring Addition and Subtraction 5
    Dragons 6
    Frog Facts 12
    Going Dotty 14
    What's My Fact? 16
    Hide and Guess 18
    Tortoise Facts 20
    Tortoise Challenges 22
    Number Lines 24
    What's the Difference? 26
    Snake It Away 28
    Double Me 31
    Grid Challenge 32
    Caterpillars 34
    Twenty Up 36
    Number Detective 38
    Join Them Up 40
    Life Rafts Puzzle 42
    Adding to 99 44
    Subtracting to 99 50
    Don't Be Vague 55
    Addition Check-Up 57
    Subtraction Check-Up 58
Exploring Multiplication 59
    Orchards 60
    What's My Story? 62
    Draw My Story 64
    Find My Story 66
    Ideas for Exploring Groups of 2 68
    Ideas for Exploring Groups of 10 69
    Ideas for Exploring Groups of 5 70
    Throw and Draw 72
    Fat Cat Facts 74
    Crocodile Facts 76
    Target Challenge 79
    Beat the Clock 81
    Multiplication Check-Up 85
Exploring Division 87
    What a Lot of Nuts 88
    More Ideas for Grouping or Sharing 91
    Division Check-Up 93
Skills Record Sheet 94
Sample Weekly Program 95
Blank Weekly Program 96

Introduction

Included in this book are many teaching ideas for adding, subtracting, multiplying, and dividing with whole numbers to 100.

To help your students work independently and think mathematically, the activities are as open-ended and flexible as possible. Unlike many teaching resources, each worksheet can be used over and over again.

Planning your number program is easy when you use the sample programs and corresponding skill guide. By exploring each topic for a week at a time, you will be able to meet the needs of at least three ability groups, with plenty to challenge your most confident students.

For example, when covering addition and subtraction, one group may be memorizing number facts to 20, while a second group is using these facts in a game. A third group may be exploring more formal sums up to 50. One group may even be exploring sums or differences to 100. The teacher's instructions are written for students too! You may have strong readers who can lead groups for you.

Just browse through the relevant section and identify the activities you want to use for that week. There are enough suggestions in each section to have up to a whole class studying each topic for at least a week!

The activities are fun, easy to implement and easy to understand. The themes have immediate appeal for young children, who will love manipulating dragons, frogs, and turtles as part of their daily math activities.

The activities are designed to maximize the way in which your students build up their knowledge of our Base 10 counting system. They are encouraged to think and work mathematically with an emphasis on mental recall and practical manipulation of objects.

A second book, *Math in Action: Numeration Activities 0–100* (TCR 3527), focuses on counting skills and place value concepts. Together these resources provide you with the practical number ideas you need to keep both you and your young students keen and motivated.

Have fun sharing the joys of operating with numbers to 100 with your students.

How to Use This Book

❏ Teaching Ideas

Each of the three units includes easy-to-implement ideas for at least one week's of teaching. Each activity includes skills and grouping strategies to help your planning, programming, and unit assessment. The skills are coded as follows:

Coded Skills (See page 94 for complete listing.)
A = Addition
S = Subtraction
M = Multiplication
D = Division

Use each idea for free exploration or guided discovery. You will never run out of ideas about what to teach your students!

❏ Student Pages

These are the basic resources you need to make the activities come alive in your classroom. There are workcards, flashcards, and spinners for games. Photocopy, color, laminate, and cut these out. Store in suitable containers with lids. Use them for free exploration by groups or individuals or for whole class demonstrations. Adapt their use for the study of numbers to 20, 50, or 100. Reuse them later for other operations.

❏ End-of-Unit Check-Ups

The activities in this book encourage open-ended exploration and independent recording. However, each unit includes a written worksheet as a check-up to see if the concepts are understood in a more formal way.

❏ Skills Record Sheet

The overall objective is to develop knowledge, skills, and understanding for the numbers 0–100 in a variety of fun, child-centered ways. Specific skills include representing number facts to 20 with objects and symbols and developing calculation procedures for operating with larger numbers beyond 20. The complete list of skills on page 94 shows you how and when these outcomes have been reached. Use this checklist to record individual responses during daily activities. Each activity lists the skills associated with it.

❏ Sample Weekly Program

The Sample Weekly Program for Mathematics on page 95 shows you one way to organize a selection of activities from the "Exploring Addition and Subtraction" unit (pages 5–58) as a five-day unit. On page 96, a blank weekly program is included.

Exploring Addition and Subtraction

In this unit, your students will do the following:

- ❏ Model real life addition/subtraction stories with objects or drawings.
- ❏ Investigate addition/subtraction combinations less than 10, to 10, doubles and to 20.
- ❏ Estimate answers to real life addition/subtraction story problems.
- ❏ Record answers using symbol/digit cards.
- ❏ Record answers using written number sentences.
- ❏ Use a number line to solve addition/subtraction problems.
- ❏ Recall addition/subtraction facts to 10 or 20 using a wide variety of strategies.
- ❏ Use addition/subtraction facts to solve two-digit problems to 99 *(optional)*.

Exploring Addition and Subtraction

Dragons

Skill
- Model, discover, and record addition/subtraction facts to 10, 20. (A1-6/S1-6)

Grouping
- Work in pairs.

Materials
- a copy of pages 7–10 for each pair colored, laminated *(optional)*, and cut as individual pieces or cards
- paper and pencils
- a timer (e.g., 5 minutes)

Directions
- Give each student a stack of ten matching dragons.
- Have them place two sets of digit and symbol cards (+, –, =) between them.
- Have them secretly take a handful of dragons and on a signal, show them to each other. Have them estimate how many of each dragon there are and how many altogether.
- Have students check by counting.
- Ask students to invent an addition or subtraction number story together, and then record their actions using the digit and symbol cards.

    e.g., "7 dragons were living in a cave. Then, 8 other dragons came to visit. That is 15 dragons altogether."

    | 7 | + | 8 | = | 1 | 5 |

- Have students write the four related number sentences on their papers.

    e.g., 7 + 8 = 15   8 + 7 = 15
    15 – 8 = 7   15 – 7 = 8

- Have students model and record as many number stories as they can in the time limit.

Variations
- Have students explore number facts to 10 by using just 10 dragons between them.
- Have students use the 0–9 spinners (page 11) to create an addition or subtraction number sentence. Have students race to model (using the dragons) and record (with digit and symbol cards) a matching dragon story together, in less than a minute.
- Challenge another team to discover as many different number facts as possible within a given time limit (e.g., 5 minutes).

Dragons   **Exploring Addition and Subtraction**

Dragons Exploring Addition and Subtraction

Dragons Exploring Addition and Subtraction

0	1	0	1
2	3	2	3
4	5	4	5
6	7	6	7
8	9	8	9

Dragons Exploring Addition and Subtraction

+	−	×	=
+	−	×	=
+	−	×	=
+	−	×	=
+	−	×	=

Dragons   **Exploring Addition and Subtraction**

Spinners 0-9

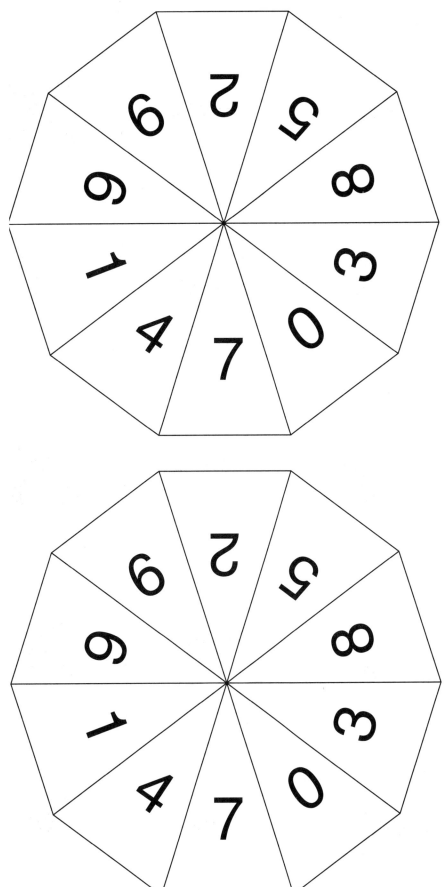

Directions: Copy and cut out each spinner. Glue or tape it to a piece of cardboard slightly larger than the spinner pattern. To use the spinner, place a large paper clip at the center of the spinner. Hold a pencil upright with the point pressed firmly at the center of the spinner, making sure the pencil point sits inside one of the ends of the paper clip. (See illustration below.) To spin the spinner, brush the free end of the paper clip with a finger.

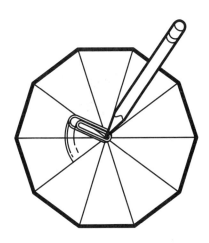

Exploring Addition and Subtraction

Frog Facts

Skill
- Model and record addition/subtraction facts about 10. (A1,3,4,5/S1,3,4,5)

Grouping
- Work in pairs.

Materials
- 10 counters (preferably small plastic frogs)
- a copy of page 13 for each pair colored, laminated *(optional)*, and cut into 10 individual cards
- two sets of 0–9 digit cards (page 9)
- one set of symbol (+, –, =) cards (page 10)
- paper and pencils
- a timer (e.g, 5 minutes)

Directions
- Have students shuffle the frog cards and place them face down between them.
- Then, have them place the digit and symbol cards face up between them.
- Have them turn over the top frog card and take turns telling each other a frog number story about their card and model this together with the counter and digit/symbol cards. (e.g., "10 frogs were sitting on the lily pad then 9 jumped into the water. That leaves just one left on the lily pad.")

| 1 | 0 | – | 9 | = | 1 |

- Have students try to interpret the story on the card in four different ways. (e.g., "Ten frogs decided to jump into the water but at the last minute one chickened out.")

| 1 | 0 | – | 1 | = | 9 |

- Have students write the four related number sentences on their papers.
  e.g., 10 – 9 = 1   10 – 1 = 9
          1 + 9 = 10   9 + 1 = 10

Variation
- Have students shuffle the cards and hold them up one by one to their partner. Can he or she tell you all four frog facts for each card in less than a minute? Exchange roles after each card.

#3528 Math in Action

Frog Facts Exploring Addition and Subtraction

Exploring Addition and Subtraction

Going Dotty

Skill
- Model, estimate, and record addition/subtraction facts to 18. (A2,5,6/S2,5,6)

Grouping
- Work in small groups.

Materials
- a set of Double Nine dominoes
- a copy of page 15 for each player and pencils
- a timer *(optional)*

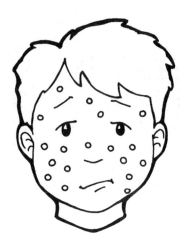

Directions
- Have students place all the dominoes face down in the center of their group.
- In turn, have them look at one of the dominoes and copy it onto their worksheets. Have them record the four number sentences to match in the space beside the drawing.

  e.g.,   8 + 9 = 17   9 + 8 = 17   17 − 9 = 8   17 − 8 = 9

Variations
- Have students explore doubles facts by playing "Double It." Call out a number from 1–9. Have students draw that many dots on the left side of one domino. Then, have them draw the same number of dots on the other side of their domino and write the matching number sentence.
- Have students play a version of "Double It" with a partner. Have students secretly draw 1–9 dots on one side of a domino and show it to their partner. Ask, "Can your partner predict the double without having to draw in the matching dots first?" Have them check by drawing in the matching dots and counting up how many altogether.
- Have students explore "doubles plus one" facts by playing "Doubles Plus One." Call out a number from 1–9. Have them draw that many dots on the left side of one domino. Then have them draw the dots for one more than this number on the other side of their domino and write the matching number sentence.

Going Dotty   Exploring Addition and Subtraction

What's My Fact?

Skill
- Model and record addition/subtraction facts to 20. (A2,5,6/S2,5,6)

Grouping
- Work in pairs.

Materials
- a copy of page 17 for each pair (cut up into 10 cards)
- paper, scissors, glue, pencils

Directions
- Have students take one of the cards and secretly circle groups to show a number fact.

   e.g.,

- Have students exchange cards with their partner and paste the cards onto their papers.
- On a signal, have them race to record their new number fact in as many different ways as they can.

   e.g.,   4 + 7 = 11 $\begin{array}{r} 4 \\ +7 \\ \hline 11 \end{array}$

- Ask, "Can you complete four cards in two minutes?"

Variation
- Have students play as a small group or do a whole class activity. One student calls out an action for everyone to follow. Each student takes a new card for each turn.

   e.g., "Show 6 and then 6 more."

   "Draw 3 and 3 and 3 and 3. How many altogether?"

   "Show 17. Now take away 7."

What's My Fact?   **Exploring Addition and Subtraction**

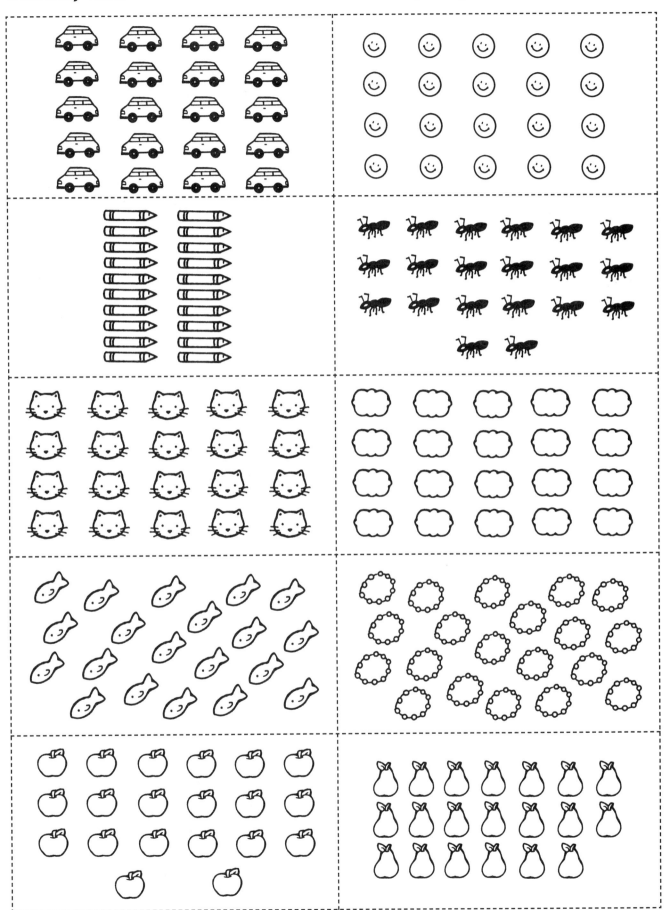

17

©Teacher Created Resources, Inc.   #3528 Math in Action

Exploring Addition and Subtraction

Hide and Guess

Skill
- Model and record + facts to 10 or 20. (A1,2,5,6)

Grouping
- Work in pairs.

Materials
- a copy of page 19 for each student (preferably on cardboard then cut into four cards)
- scissors and colored pencils
- paper for recording discoveries

Set-Up
- Crease along the dashed fold line. Cut along the two lines as shown (a).

a.

- Close the last two sections as shown (b). Draw 0–10 objects in the first section.

b.

- Close the first and third sections as shown (c). Draw 0–10 objects in the middle section.

c.

- Count how many objects you have drawn. Close the first and second sections as shown (d). Write the matching total in the last section. You are now ready to start Hide and Guess.

d.

Directions
- Have students close one section and show it to their partner. Can he or she guess the hidden number in less than 10 seconds?
- Have students record their turn by writing their facts horizontally or vertically.

Variation
- Have students find a way to show subtraction facts using a blank Hide and Guess card.

#3528 Math in Action
18
©Teacher Created Resources, Inc.

Exploring Addition and Subtraction

Tortoise Facts

Skill
- Recall addition/subtraction facts to 10 or 20. (A8,9/S8,9)

Grouping
- Work in small even-numbered groups.

Materials
- a copy of page 21 for each student (preferably on cardstock, cut up into 10 cards, with a different color for each team)
- pencils

Directions
- Each student in a team selects a different number from 0–10. (Let's say 8.)
- Have students record up to 10 addition facts about their number on the blank tortoise cards.

 e.g.,

- Have students shuffle all the cards together and deal them out to each player.
- Tell students to turn and face a partner. Have one student in each pair hold up a card and ask their partner to say the complete number fact.

e.g., Nine plus one is ten.

- Have students exchange roles.

Variations
- Have students sort all the tortoise cards quickly into same number fact piles.
- Write tortoise fact cards up to 10 + 10.
- Write subtraction fact cards to 10 or 20. These could be in a separate color.
- Ask, "How many different tortoise fact cards can you discover for each number?"
- Have a group challenge: Name all the tortoise facts in the shortest time.
- Play Tortoise Snap as follows: Deal cards face down to each player. Each player places a card face up on a center pile. Call out "Snap" when he or she sees two matching tortoise fact cards. Win all the cards in the central pile at that time.

Tortoise Facts Exploring Addition and Subtraction

Exploring Addition and Subtraction

Tortoise Challenges

Skills
- Recall addition facts to 10 or 20. (A8,9)
- Recall subtraction facts to 10 or 20. (S8,9)

Grouping
- Work as individuals up to the whole class.

Materials
- a copy of page 23 for each student
- pencils
- a one-minute timer *(optional)*

Directions
- Have each student write a random number from 0–10 and a plus sign in the center of a tortoise's shell as shown (a).

a.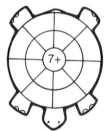

- Have each student write eight random numbers from 0–10 in the middle section of the tortoise's shell as shown (b).

b.

- Have each student fill in the outer shell by adding the center number to each middle number as shown (c).

c.

- Have students time themselves. Ask, "Can you complete each Tortoise Challenge in less than one minute?"
- Tell students, "Once you have completed all nine tortoises, find your fastest time. Are some numbers easier to add than others?"

Variation
- Have students practice subtraction facts by writing a random number from 10–20 and a subtraction sign in the center and then random numbers from 0–10 in the middle section.

Tortoise Challenges   **Exploring Addition and Subtraction**

Number Lines

Skills
- ❑ Use a number line to solve addition problems. (A7)
- ❑ Use a number line to solve subtraction problems. (S7)

Grouping
- ❑ Work in groups of up to four students.

Materials
- ❑ one number line for each student (laminate page 25 and cut into four pieces) The children number lines can be used as 0–10 or joined to make 0–20.
- ❑ a set of 0–9 digit cards (page 9)

Directions
- ❑ Have students shuffle the digit cards and place them face down in the center of their group.
- ❑ In turn, have them turn over one card to find their starting point (e.g., 9). Have them find this point on their number line.
- ❑ Next, have them turn over a second card (e.g., 8). This shows how many to add to their first number.
- ❑ First, have them estimate where they will land. Then have them count by moving their finger from the first number to check.

Variations
- ❑ Silent Count
  A leader turns over a card to identify everyone's starting point (e.g., 6). The leader then claps the number shown on a second card (e.g., 8) while the rest of the class counts silently. On a signal, everyone calls out their final total (e.g., 14).
- ❑ Subtract It
  Have a student call out a random number from 10–20 and find this point on his or her number line. Next, have him or her turn over a digit card to see how many to take away.

Number Lines   **Exploring Addition and Subtraction**

25

Exploring Addition and Subtraction

What's the Difference?

Skill
- Create and solve subtraction story problems. (S3)

Grouping
- Work in pairs.

Materials
- page 27 (cut into 10 cards)
- 20 counters
- paper and pencils

Directions
- Have students place all the story cards face down in a pile in the center of their group. Have them turn over the top card.

    e.g.,
    > Ombie takes
    > 6 minutes
    > to walk to school.
    > Julien takes 13 minutes.

- Have students discuss each story in terms of who has more, who has less.
    e.g., "13 is more than 6."
    "6 is less than 13."

- Have students match the story using counters and record this on their papers.
    e.g., "13 is 7 more than 6."
    "13 minutes is 7 more minutes than 6 minutes."
    "6 is 7 less than 13."

- Ask, "What's the difference between the two numbers?"
    e.g., "The difference between 13 and 6 is 7."

Variation
- Have students write some more "What's the Difference?" problems for another pair to solve. Have them exchange problems. Ask, "How many can you do in five minutes?"

What's the Difference?   **Exploring Addition and Subtraction**

Sam has 5 horses. Maria has 8 horses.	My apartment is on the 14th floor. Your apartment is on the 8th floor.
My tank has 7 fish. Your tank has 9 fish.	Lucy has 8 friends. Alex has 17 friends.
I have $15. You have $6.	Ombie takes 6 minutes to walk to school. Julien takes 13 minutes.
Bill won 6 games of Snap. Di won 11 games.	Minh's backyard is 16 yards long. Ali's backyard is 6 yards long.
Taso drank 5 cartons of milk last week. Amelia drank 9 cartons of milk.	Maggie has 8 favorite TV programs. Tom has 8 favorite TV programs too.

Exploring Addition and Subtraction

Snake It Away

Skill
- ❏ Recall subtraction facts to 10, 20 using a range of strategies. (S8,9)

Grouping
- ❏ Work in groups of two to six players.

Materials
- ❏ a Snake It Away strip for each player (page 29 cut into six strips)
- ❏ Snake It Away cards (page 30 laminated, cut into pack of 20 cards)
- ❏ pencils
- ❏ paper (optional)

Directions
- ❏ Have students shuffle the cards face down in a pile in the center of their group. Have them turn over the top card.
- ❏ Each player crosses off two numbers on his or her snake, which have that difference.

e.g.,  Cross out 18 and 5,
or 14 and 1,
or 13 and 0.

- ❏ In turn, have them say their matching number sentence aloud.

  e.g., "18 take away 5 is 13."

  "The difference between 18 and 5 is 13."

- ❏ The first player to "snake away" (cross off) all the numbers from 1–20 is the winner.

Variations
- ❏ For a simpler game, have students redraw the worksheet with numbers from 0–10 on each snake.
- ❏ Have students cross off two numbers which add to the number shown on the card.
- ❏ Have students record their number sentence on their paper at the end of their turn each time.

Snake It Away Exploring Addition and Subtraction

Snake It Away Exploring Addition and Subtraction

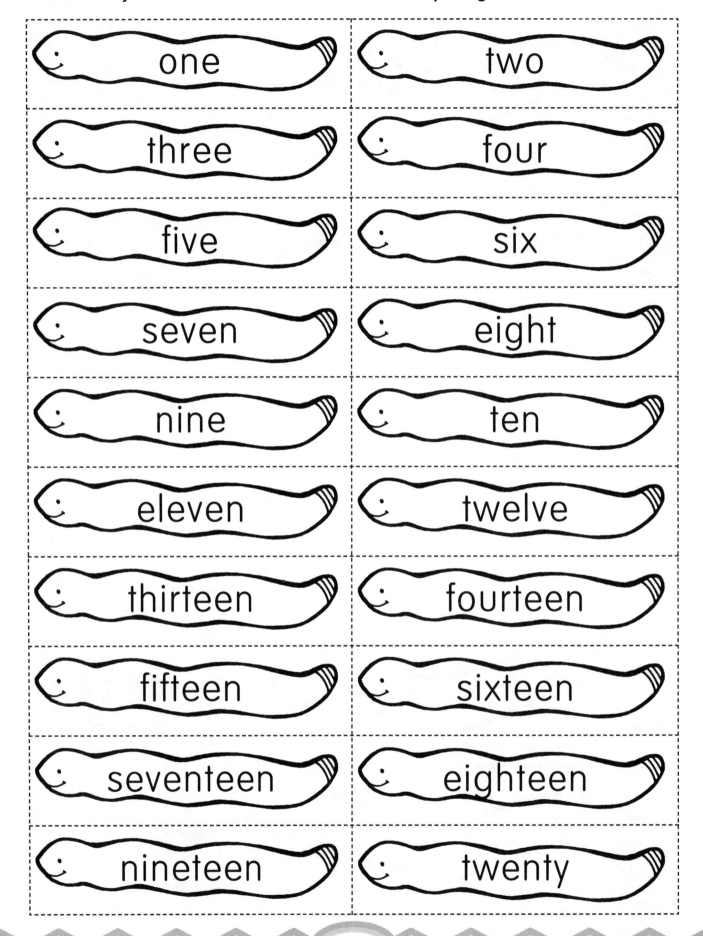

30

Exploring Addition and Subtraction

Double Me

Skill
- Recall addition facts to 20 using a variety of strategies. (A9)

Grouping
- Work in pairs.

Materials
- a set of 0–9 digit cards (page 9)
- 20 counters, a 0–20 number line (page 25), a calculator
- a timer *(optional)*

Directions
- Double Me
  Have students shuffle the cards and place them face down in the center. One student turns over a card. Have him or her race to identify the double of that number before his or her partner does. He or she can use any method to get the answer (e.g., mental math, counters, a number line, a calculator). Switch roles.

- Double Me Plus One
  Have a student turn over a card. He or she must identify double that number plus one before his or her partner does. He or she can use any method to get the answer.

- Double Me Less One
  Have a student turn over a card. He or she must identify one less than double that number before his or her partner does. He or she can use any method to get the answer.

- Double Me Mind Muncher Challenge
  State which challenge (e.g., Double me Plus One) each student wants his or her partner to try. This student will shuffle the cards and hold them. Have each student face his or her partner and work through all ten cards calling out each one to his or her partner who tries to say each answer as quickly as possible. Have students exchange roles and try to improve the speed when they play this again.

Exploring Addition and Subtraction

Grid Challenge

Skill
- ❑ Recall addition facts to 10, 20 using a range of strategies. (A8,9)

Grouping
- ❑ Work as individuals, small groups, or up to the whole class.

Materials
- ❑ Grid Challenge strip for each player (page 33 cut into four strips)
- ❑ pencils
- ❑ calculator *(optional)*
- ❑ math workbooks *(optional)*

Directions

Have students follow the directions below.
- ❑ Write four random digits from 0–9 in the four spaces in each grid square as shown (a).

- ❑ Next, add across to find the totals. Write the totals in the matching circles as shown (b).

- ❑ Then add down to find the next totals. Write the new totals in the matching circles as shown (c).

- ❑ Finally, add across the two lower circles to find the total for the star as shown (d). Check by adding the two right-hand circles. Use a calculator to help if necessary.

a.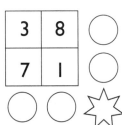

b.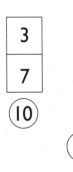

c.

d.

Variations
- ❑ Each player writes the same four random digits in each square. On a signal, have students race other players in their team to record the answers. Have them try to be the first to complete all three grids.
- ❑ Use these grids later as multiplication grids. Use a calculator to find the star answers.
- ❑ Have students cut out and paste grids into their workbooks.

Grid Challenge **Exploring Addition and Subtraction**

Exploring Addition and Subtraction

Caterpillars

Skills
- Add three or more digits up to 20. (A10)
- Subtract three or more digits from up to 20. (S10)

Groupings
- Work in small groups or whole class.

Materials
- caterpillars for each player (page 35 cut into 16 pieces)
- one die
- pencils, glue, paper
- calculator

Directions
- Have students throw the die four times and write the digits in the four spaces on a caterpillar.

- On a signal, have students race to add up all four digits.
- Have students check their totals using calculators and then glue their caterpillars onto blank paper. Have them write their totals beside the caterpillars.

Variations
- Ask, "How many caterpillars can you complete in five minutes?"
- For a simpler version, have students just throw the die three times and record three numbers on their caterpillars.
- Take It Away
  Have students play a subtraction version by starting at 20 and subtracting the four digits mentally. Have them check using a calculator.
- Caterpillar Challenge
  Have students try to be the first in their teams to add or subtract the four numbers in their heads.

Caterpillars   **Exploring Addition and Subtraction**

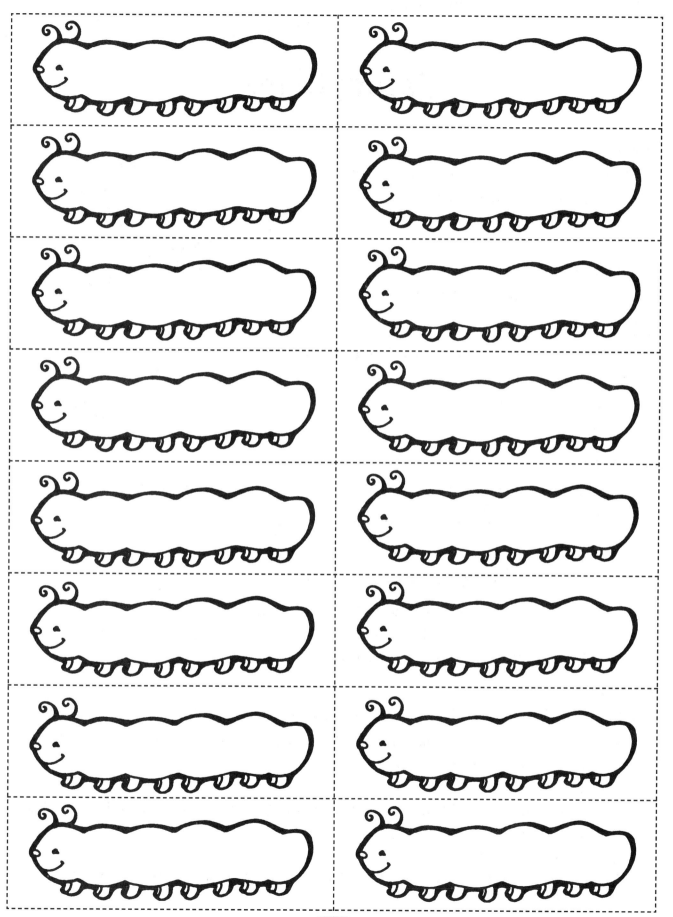

35

Exploring Addition and Subtraction

Twenty Up

Skill
- Recall addition facts to 20 using a variety of strategies. (A9)

Grouping
- Work in pairs.

Materials
- a set of 0–9 digit cards for each player (page 9)
- a Twenty Up grid (page 37)
- pencil, paper to record scores

Directions
- Have students take turns placing a card anywhere on the grid.
- Tell students to score 10 points every time they make a line that adds exactly to 20. A line can be any three or four cards in a row, column, or diagonal. There can be a space between three cards.

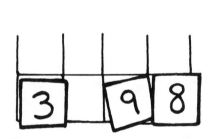 or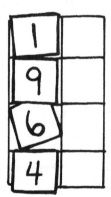

- Have students keep playing by placing cards in turn until all the spaces have been filled.
- Tell students to try to be the player with the highest score at the end of the game.

Variations
- Reverse the rule. If a line adds to 20, you lose 10 points.
- Select a different total (e.g., 17) for each round.

Exploring Addition and Subtraction

Number Detective

Skills
- Recall addition facts to 20 using a variety of strategies. (A9)
- Recall subtraction facts to 20 using a variety of strategies. (S9)

Grouping
- Work in pairs.

Materials
- 20 counters (e.g., tiny teddies, dinosaurs)
- two empty margarine containers with lids
- "Number Detective" (page 39 cut into four pieces)
- pencils

Directions
- Have a student secretly place some counters in two containers. Have him or her count up how many altogether but keep the total secret.
- Have that student put the lid on each container and then place them in a line near his or her partner. Say, "Ask your partner whether they would like to see inside the left or the right container."
- Have the student show the partner the contents of one container. Then have the student tell the partner how many counters there are altogether in both containers.
- Say, "Ask your partner to be a Number Detective and guess how many counters must be hidden inside the other container."
- Have students record their actions on the worksheet.

 + | 8 | = | 15 |

- Have students check by finding how many counters were hidden.

Variation
- For subtraction, have a student take some counters and let his or her partner count them. Have the first student hide some counters and show the remainder to the partner who tries to guess how many must be hidden. Have students record their actions.

| 19 | − = | 9 |

Number Detective

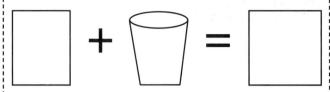
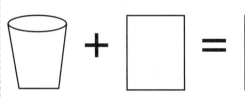
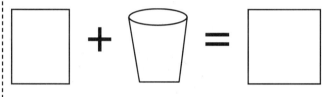
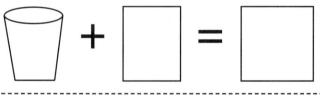

Number Detective

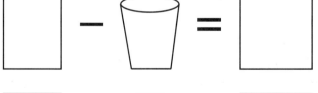
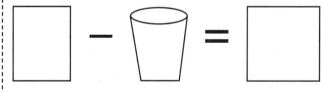
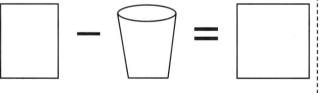

Number Detective

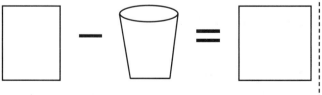

Exploring Addition and Subtraction

Join Them Up

Skill
- ❏ Recall addition facts to 10, 20 using a variety of strategies. (A8,9)

Grouping
- ❏ Work as individuals or pairs.

Materials
- ❏ a Join Them Up worksheet (page 41 cut into four pieces) for each player
- ❏ a set of 0–9 digit cards (page 9)
- ❏ pencils

Directions
- ❏ The leader calls out a number between 6 and 14 (e.g., 9).
- ❏ Each student tries to rearrange his or her digit cards in a similar position to the grid to make lines of three digits that add to the selected number.

      9   3 + 1 + 5

- ❏ Once students discover a complete solution, have them record it on their grid worksheets.

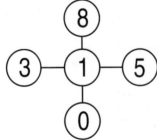

- ❏ Have students try to discover at least three solutions for each number.

            5   8   4
  9   2 3 4   3 1 5   0 2 7
            1   0   3

Variation
- ❏ Use a timer. Have students discover one solution in less than one minute.

Join Them Up

Exploring Addition and Subtraction

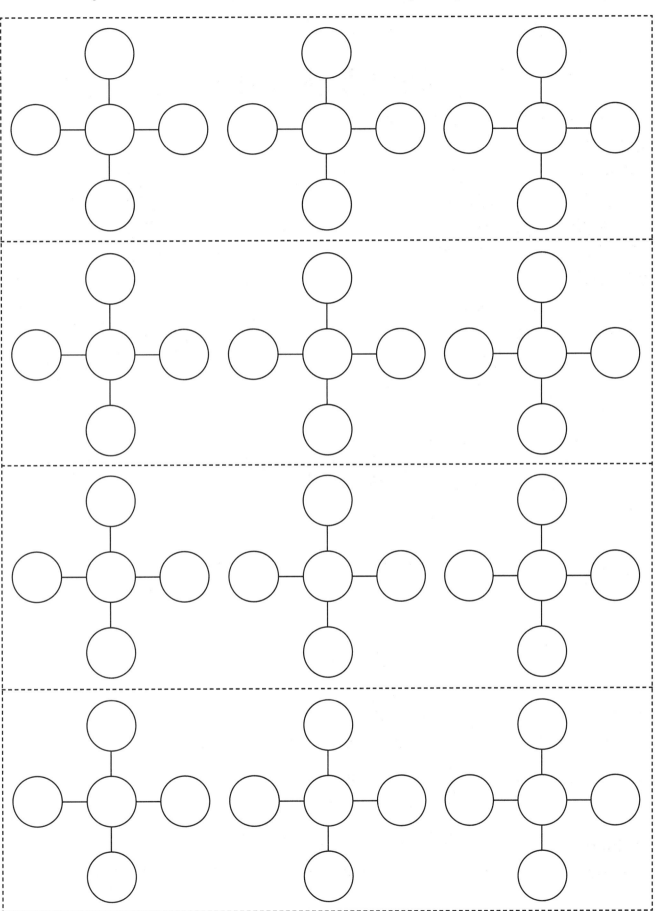

Life Rafts Puzzle

Skill
- Recall addition facts to 10, 20 using a variety of strategies. (A8,9)

Grouping
- Work in small groups of two to three students.

Materials for Each Team
- Set A: Life Rafts clues (page 43 cut into five pieces)
- three objects to be the Life Rafts (e.g., sheets of paper, small boxes)
- 10 counters to be the people
- scrap paper
- pencils
- paper

Directions
- Tell students, "You are ready for a Super Addition Challenge. Let's try this puzzle."
- Tell them to look at the five clues together and decide what the problem is about. Ask, "How many people are there? How many rafts do you need?"
- Tell students to model their problem with the counters and "rafts." Tell them to use their clues to work out how many people were placed on each raft.
- Once they discover a solution, tell them to find a way to record this on their papers.

Variations
- Ask students, "How many other ways can you find to arrange this many people between the rafts?"
- Use Set B: Life Rafts clues (page 43 cut into five pieces) with 20 counters.
- Have students make up their own Life Rafts clues for another team to solve (e.g., use 16 people and 4 rafts).

SET A: Life Rafts

Ten people jump off a sinking ship.

There are three life rafts.

The first and the third rafts together have six people.

The third raft has more than one person.

The first and second rafts together have seven people.

SET B: Life Rafts

Twenty people jump off a sinking ship.

There are three life rafts.

The first and the third rafts together have thirteen people.

The second and third rafts together have twelve people.

The first raft has fewer than ten people.

Exploring Addition and Subtraction

Adding to 99

Skills
- ❏ Model addition to 99 using objects. (A11)
- ❏ Record addition to 99 without trading. (A12)
- ❏ Record addition to 99 with trading *(optional)*. (A13)

Grouping
- ❏ Work in small groups.

Materials
- ❏ place value boards (page 45)
- ❏ place value materials (e.g., beansticks, pasta sticks, craft sticks, matchsticks, beads)
- ❏ Adding to 99 activity cards (pages 46–49, each cut into two cards)
- ❏ a set of 0–9 digit cards (page 9) or a 0–9 spinner (page 11)
- ❏ a hundreds chart
- ❏ scrap paper, pencils, math workbooks
- ❏ calculators
- ❏ timers

Directions
- ❏ Tell students, "You're now ready to extend your skills at adding to numbers beyond 20."
- ❏ Have students work through the activity cards in sequential order starting with the first card labeled with one dot in the bottom right corner and ending with the last card labeled with eight dots. Tell them to remember to use their place value language. (e.g., 36 is thirty-six. That is three tens and six ones.)
- ❏ Remind them that when adding two numbers, start with the ones digit first. Then place the objects on the place value board in the matching positions. Tell them to find out how many ones altogether. Are there enough ones to make another ten?
- ❏ Next, tell them to find out how many tens.
- ❏ When recording, have students write the numbers under each other as tens and ones.

Variation
- ❏ For an easier activity, explore adding to numbers up to 50.

Adding to 99 *Exploring Addition and Subtraction*

Tens	Ones

Adding to 99 — Exploring Addition and Subtraction

Trade Up a Ten

✔ The leader calls out a number between 20 and 90.

✔ Model this with materials on your place value board. Each player then turns over a digit card. Add this many to your first number.

✔ Remember to trade ten ones for one ten if you collect enough ones. Record your actions in your workbook.

e.g.,
```
  45          73
+  4        +  9
  49          82
```

How Many More to the Next Ten?

✔ Work with a partner. Call out any number between 0 and 99.

✔ Ask your partner to quickly tell you how many more you need to add to make the next ten.

e.g., 36 You need 4 more to make 40.
52 You need 8 more to make 60.
89 You need 1 more to make 90.

✔ Check with a calculator. Exchange roles.

✔ How many questions like this can you answer in one minute?

How Many Tens?

- ✔ Work with a partner. Use the 0–9 spinner twice to tell you how many tens to add.

- ✔ Race your partner to find out how many tens altogether.

- ✔ Record these in your workbook.

   e.g.,
   ```
        9 tens        90
      + 6 tens      + 60
       15 tens       150
   ```

- ✔ How many can you do in one minute?

Look For a Pattern

- ✔ Use a hundreds chart to help you. Watch what happens when you keep adding the same number.

   e.g., 8 + 7

   18 + 7   Can you guess the next number?

   28 + 7   What's the pattern?

- ✔ Explore what happens with other number facts too.

   e.g., 6 + 5   9 + 7   2 + 3

   16 + 5   19 + 7   12 + 3

   26 + 5   29 + 7   22 + 3

- ✔ Can you guess the next number each time?

Adding to 99 — Exploring Addition and Subtraction

Mental Mind Muncher

- ✔ Try adding one-digit numbers to two-digit numbers in your head.

- ✔ Your partner calls out a number between 20 and 90 and a second number between 0 and 9.

- ✔ Can you add these two numbers mentally?

- ✔ Check with your calculator.

- ✔ Record your successes in your workbook.

Race Me

- ✔ You need four people on your team. One person uses objects, one uses pencil and paper, one uses a calculator, and one uses mental skills.

- ✔ The leader calls out a number between 20 and 90, and then turns over a digit card. On a signal, the leader calls out "Add." Race your teammates to find the answer.

- ✔ Exchange roles regularly.

- ✔ Which method works best for you?

Estimate It!

- ✓ You will need a calculator and a partner.

- ✓ On a signal, you both call out a number between 20 and 50.

- ✓ Estimate what the total will be when you add these numbers.

- ✓ Tell your partner your estimation then check with your calculator. Were you close?

Add These

- ✓ You will need two sets of digit cards, a calculator, and a partner.

- ✓ On a signal, call out a number between 50 and 100 (e.g., 80).

- ✓ Race your partner to discover pairs of cards that add exactly to that number.

    e.g.,   80   47
                    + 33

- ✓ How many can you discover in three minutes? Record your favorite ones in your workbook.

- ✓ For a super challenge, just use one set of digit cards.

Exploring Addition and Subtraction

Subtracting to 99

Skills
- ❑ Model subtraction to 99 using objects. (S11)
- ❑ Record subtraction to 99 without trading. (S12)
- ❑ Record subtraction to 99 with trading *(optional)*. (S13)

Grouping
- ❑ Work in small groups.

Materials
- ❑ place value boards (page 45)
- ❑ Subtracting to 99 activity cards (pages 51–54, each cut into two cards)
- ❑ place value materials (e.g., beansticks, pasta sticks)
- ❑ a set of 0–9 digit cards (page 9) or a 0–9 spinner (page 11)
- ❑ a hundreds chart
- ❑ scrap paper and pencils
- ❑ calculators

Directions
- ❑ Say to students, "You are now ready to extend your skills at subtracting to numbers beyond 20."
- ❑ Have students work through the activity cards in sequential order starting with the first card labeled with one dot in the top right corner and ending with the last card labeled with eight dots.
- ❑ Tell them that when subtracting, model the largest number only. Tell them to remember to use place value language too.

    e.g., 73 is seventy-three. That is seven tens and three ones.

    Tell students to place the objects on the place value board in the matching positions.
- ❑ Next, tell them to start with the ones digit. Ask, "Do you have enough ones? Remember that you can trade one ten for ten ones."
- ❑ Say, "When you discover your answer, record the numbers under each other as tens and ones.

Variation
- ❑ For an easier activity, explore subtraction to 50.

Subtracting to 99 — *Exploring Addition and Subtraction*

Trade Down a Ten

✔ The leader calls out a number between 30 and 99.

✔ Model this with materials on your place value board.

✔ Each player then turns over a digit card. Take this many away from your first number. Remember you can trade one ten for ten ones if there are not enough ones.

✔ Record your actions on your paper.

e.g.,
```
  5 5        8 6
-   3      -   7
-----      -----
  5 2        7 9
```

How Many Do I Remove?

✔ Work with a partner. Call out any number between 0 and 99.

✔ Ask your partner to quickly tell you how many you need to take away to get the next ten.

e.g., **47** You need to remove 7 to make 40.
61 You need to remove 1 to make 60.
39 You need to remove 9 to make 30.

✔ Exchange roles. How many questions like this can you answer in one minute?

Subtracting to 99 — *Exploring Addition and Subtraction*

How Many Tens?

- ✔ Work with a partner. Use the 0–9 spinner twice to tell you how many tens to subtract. Remember that the largest number goes first.

- ✔ Race your partner to find out how many tens are left.

- ✔ Record these on your paper.

    e.g.,
    ```
      6 tens      6 0
    - 5 tens    - 5 0
    ───────    ──────
      1 ten       1 0
    ```

- ✔ How many can you do in one minute?

Do I Need to Trade?

- ✔ Work with a partner. Ask your partner to call out a number between 30 and 100 and then another number from 0 to 9.

- ✔ Do you need to trade to remove the smaller number? Tell your partner your idea, then find a way to check.

- ✔ Record these on your paper in two columns.

    e.g.,

Trade	No Trade

Subtracting to 99 — *Exploring Addition and Subtraction*

Look For a Pattern

- ✔ Use a hundreds chart to help you. Watch what happens when you keep taking the same number away.

    e.g.,   7 – 5

           17 – 5   Can you guess the next number?

           27 – 5   What's the pattern?

- ✔ Explore what happens with other number facts too.

    e.g.,   16 – 9   14 – 7   12 – 3

           26 – 9   24 – 7   22 – 3

           36 – 9   34 – 7   32 – 3

- ✔ Can you guess the next number each time?

Mental Mind Muncher

- ✔ Try subtracting one-digit numbers from two-digit numbers in your head.

- ✔ Your partner calls out a number between 20 and 99 and a second number between 0 and 9. Can you work out how to take away the smaller number in your head?

- ✔ Check with your calculator.

- ✔ Record your successes on your paper.

Subtracting to 99   **Exploring Addition and Subtraction**

Race Me

- ✔ You need four people on your team. One person uses objects, one uses pencil and paper, one uses a calculator, and one uses mental skills.

- ✔ The leader calls out a number between 20 and 90, then turns over a digit card. On a signal, the leader then calls out "Take it away." Race your teammates to find the answer.

- ✔ Exchange roles regularly.

- ✔ Which method works best for you?

Estimate It

- ✔ You will need a calculator and a partner.

- ✔ On a signal, you both call out a number between 50 and 100. Estimate what the difference will be when you take the smaller number away from the larger number.

- ✔ Tell your partner your estimation and then check with your calculator.

- ✔ Were you close?

Exploring Addition and Subtraction

Don't Be Vague

Skills
- ❏ Model addition or subtraction to 99 with objects. (A11/S11)
- ❏ Record addition and subtraction to 99 with/without trading. (A12,13/S12,13)

Grouping
- ❏ Work in small groups.

Materials for Each Team
- ❏ place value boards (page 45)
- ❏ place value materials (e.g., beansticks, pasta sticks, craft sticks, toothpicks, beads)
- ❏ Don't Be Vague activity cards (page 56 cut into 10 cards)
- ❏ scrap paper and pencils
- ❏ calculators

Directions
- ❏ Tell students, "Sometimes people can be quite vague or not clear. The activity cards you will be reading will have math problems that are vague or not clear. Look at one of the activity cards. Discuss with your team the different possibilities for answers."

    e.g.,   A number between 30 and 40 could be 32, 37, 38.
            A number between 20 and 30 could be 25, 29, 21.

- ❏ Say, "As a group, try to find all the different ways each problem on the activity card can be solved."

    e.g.,   34 – 22, 39 – 21, 36 – 29

- ❏ Have students use the place value boards and materials to help them work out each answer.
- ❏ Once they have discovered a solution, have students copy the problem on their papers and record their answers.
- ❏ Ask, "How many different answers can you discover for each problem?"

Variation
- ❏ Have students make up Don't Be Vague cards for another team to solve.

Don't Be Vague — **Exploring Addition and Subtraction**

Sam and Lucy collect fish. Sam has between 30 and 40 fish. Lucy has between 20 and 30 fish.

How many fish altogether?

Taso had more than 50 sheep but fewer than 60 sheep. Yesterday he sold 17 sheep.

How many sheep does he have now?

Maha has $70 to $80 in the bank. Ravi has $10 to $20 in the bank.

How much money is this altogether?

Max hates peas. He had between 40 and 50 on his plate. He secretly gave some to his dog Sasha. There are now 10 peas.

How many peas did Sasha eat?

Ernie ate between 10 and 20 mangoes last week. They were delicious.

Cathy ate between 10 and 20 mangoes, too.

How many mangoes is that?

Bridget loves tadpoles. She had a big jar with between 70 and 80 in it. She gave some to her friend Tom. Bridget now has only 40 tadpoles.

How many does Tom have?

My hens laid more than 50 but fewer than 60 eggs this week. Your hens laid between 20 and 30 eggs.

How many eggs altogether?

Bill reads books. He has between 60 and 70 books. Di reads books, too. She has fewer than 30 but more than 20 books.

How many more books does she need to have as many as Bill?

There are two aliens. One alien has between 40 and 50 legs. The other alien has 30 to 40 legs.

How many legs altogether?

Ming, our farm cat, catches mice. Last week she caught more than 90 but less than 100. This week she caught fewer than 35 but more than 25.

What's the difference?

Exploring Addition and Subtraction

Addition Check-Up

1. Draw dots to show four different ways to make a domino add to 9.

2. Make each pair add to 10.

3. Write your own story about 6 plus 7.

4. Answer these facts as quickly as you can.

   a. 0 + 6 = ☐   b. 10 + 4 = ☐   c. 2 + 9 = ☐   d. Double 9 ☐

5. Be a number detective and find these missing numbers.

   a. 3 + ☐ = 12   b. ☐ + 2 = 9   c. 4 + ☐ = 11   d. ☐ + 6 = 14

   e. 10 + ☐ = 13   f. ☐ + 7 = 16   g. 2 + ☐ = 8   h. ☐ + 10 = 15

6. Help work out how many oranges were sold each week day.

   a. 44   36   17   52   41
       + 2   +23   +31   +46   +58

   b. 39   64   74   57   23
       + 5   +18   +26   +34   +49

Exploring Addition and Subtraction

Subtraction Check-Up

1. Cross out dragons to show 14 − 8 =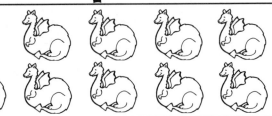

2. Write your own story about 16 take away 7.

3. Join the matching facts.

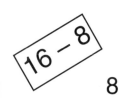 9   6    11   3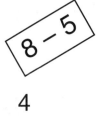
        8   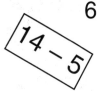 5   10   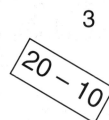 4

4. Answer these facts as quickly as you can.

a. 10 − 4 = ☐   b. 8 − 3 = ☐   c. 17 − 8 = ☐   d. 13 − 6 = ☐

5. Be a number detective and find these missing numbers.

a. 14 − ☐ = 10   b. ☐ − 3 = 7   c. 15 − ☐ = 9   d. ☐ − 5 = 8

e. ☐ − 2 = 9   f. 16 − ☐ = 8   g. ☐ − 5 = 7   h. 16 − ☐ = 6

6. Help work out how many fish are left in each tank.

a.   23   37   46   58   69
    − 2   −15   −23   −33   −46
    ___   ___   ___   ___   ___

b.   28   56   61   74   83
    − 9   −17   −28   −39   −46
    ___   ___   ___   ___   ___

Exploring Multiplication

In this unit, your students will do the following:
- ❑ Model and draw multiplication as equal groups or rows.
- ❑ Create and solve story problems.
- ❑ Estimate answers to story problems.
- ❑ Record using symbol cards (e.g., x, =) and written number sentences.
- ❑ Recall x2, x10, x5 and 2x (doubles) facts.
- ❑ Model, draw, record, recall x1, x0, x3, x4 facts *(optional)*.

Exploring Multiplication

Orchards

Skills
- ❏ Model multiplication as equal groups. (M1)
- ❏ Use multiplication language: "groups of" or "makes." (M2)
- ❏ Create and solve story problems. (M3)
- ❏ Record using symbol and digit cards. (M4)

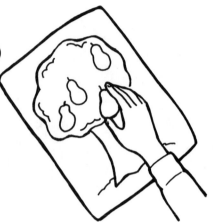

Grouping
- ❏ Work in groups of up to five students.

Materials for Each Group
- ❏ 10 copies of the fruit tree (page 61 photocopied, decorated, and laminated)
- ❏ 100 counters (e.g., plastic fruit or bottle tops)
- ❏ two sets of digit cards (page 9)
- ❏ symbol cards (x, =) (page 10)
- ❏ paper
- ❏ pencils for recording

Directions
- ❏ Have students work as a team. Have them decide how many trees in the orchard for this round and how many pieces of fruit will be on each tree.
- ❏ Have students estimate how much fruit this will be altogether. Have them discuss their estimations and then check by modelling with the counters and trees.
    e.g., 4 trees, 3 pieces of fruit
- ❏ The student with the closest estimation scores two points.
- ❏ Have students record the story using the symbol and digit cards, and then copy it on their papers. Have students draw a picture to match.
- ❏ Ask, "How many orchard stories can your team solve in five minutes?"

Variation
- ❏ Have students challenge the rest of their team to solve their problem.
    e.g., There were 20 pieces of fruit altogether. How many trees would be in the orchard if each tree had the same amount?

Orchards — Exploring Multiplication

Exploring Multiplication

What's My Story?

Skills

- ❏ Model multiplication as equal groups or rows. (M1)
- ❏ Use multiplication language: "groups of," "rows of," or "makes." (M2)
- ❏ Create and solve story problems. (M3)
- ❏ Record using symbol and digit cards. (M4)

Grouping

- ❏ Work in groups of up to five students.

Materials for Each Group

- ❏ What's My Story? cards (page 63 photocopied, laminated, cut up into 20 cards)
- ❏ 100 counters (e.g., plastic frogs, dinosaurs, clowns, sea animals, cubes)
- ❏ two sets of digit cards (page 9)
- ❏ symbol cards (x, =) (page 10)
- ❏ paper and pencils for recording

Directions

- ❏ Have students shuffle the cards and place them face down in the center of their group. Have one person select a card, and then create and model a story based on this card.

e.g.,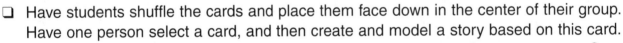

"There were 2 ponds. Inside each pond there were 6 frogs. How many frogs altogether?"

e.g.,

"There were 2 lilypads. On each lilypad there were 6 frogs sitting in a line. How many frogs altogether?"

- ❏ Each student records his or her actions in the story and asks the group to check.

Variations

- ❏ Have students select the best multiplication story to add to a large class display book.
- ❏ Have students make their own set of cards for another team to use.

What's My Story? Exploring Multiplication

3 groups of 2	1 row of 5	2 groups of 10	2 groups of 1
6 groups of 2	3 rows of 5	4 groups of 10	2 groups of 4
10 groups of 2	5 rows of 5	6 groups of 10	2 groups of 6
4 groups of 2	8 rows of 5	8 groups of 10	2 groups of 7
7 groups of 2	9 rows of 5	10 groups of 10	2 groups of 9

Exploring Multiplication

Draw My Story

Skills

- ❏ Model multiplication as equal groups or rows. (M1)
- ❏ Create and solve story problems. (M3)
- ❏ Estimate answers to story problems. (M6)
- ❏ Record using written number sentences. (M5)

Grouping

- ❏ Work in groups of up to five students.

Materials for Each Group

- ❏ Draw My Story cards (page 65 photocopied, laminated, and cut into 20 cards)
- ❏ paper
- ❏ pencils for recording

Directions

- ❏ Have students shuffle the cards and place them face down in the center of their group.
- ❏ A student selects a card. Each student creates and draws a story based on this card and works out the answer altogether.

    e.g., | 3 x 10 |

    I had 3 aquariums.
    There were 10 fish in each aquarium.
    I had 30 fish altogether.

- ❏ Have students discuss the different stories at the end of each round. Ask, "How many groups were there? How many in each group? How many altogether?"
- ❏ Have students record their stories as number sentences under the drawing on their papers.

    e.g., 3 x 10 = 3 0

Variations

- ❏ Have each student show a partner his or her drawing. Have him or her ask the partner to predict the number sentence.
- ❏ Have students make their own set of cards for another team to use.

4 x 2	1 x 10	3 x 5	2 x 6
7 x 2	5 x 10	8 x 5	2 x 1
10 x 2	3 x 10	1 x 5	2 x 9
2 x 2	7 x 10	6 x 5	2 x 7
5 x 2	9 x 10	4 x 5	2 x 8

Find My Story

Skills
- ❑ Model multiplication as equal groups or rows. (M1)
- ❑ Create and solve story problems. (M3)
- ❑ Estimate answers to story problems. (M6)
- ❑ Record using written number sentences. (M5)

Grouping
- ❑ Work in groups of up to four students.

Materials for Each Group
- ❑ Find My Story strips for each player (page 67 photocopied and cut up into four strips)
- ❑ paper
- ❑ pencils for recording

Directions
- ❑ Have students take turns being the leader. The leader decides on a group of 2, 5, or 10, and then calls out, "Find out how many groups of . . ." (e.g., 5).
- ❑ Each student estimates how many groups they will find and then circles the matching groups on a story strip to check. Each student can have a different strip (e.g., turtles, frogs, birds, caterpillars).
- ❑ Each player then tells the rest of the team their story.

    e.g., "I had 20 turtles. I estimated 4 groups and I was right.

    I found 4 groups of 5 turtles."

- ❑ Have students discuss the different stories at the end of each round. Ask, "How many groups were there? How many in each group? How many altogether?"
- ❑ Have students record their stories as number sentences on their papers.

    e.g., 4 x 5 = 20

Variation
- ❑ Have students make their own set of strips for another team to use.

Find My Story Exploring Multiplication

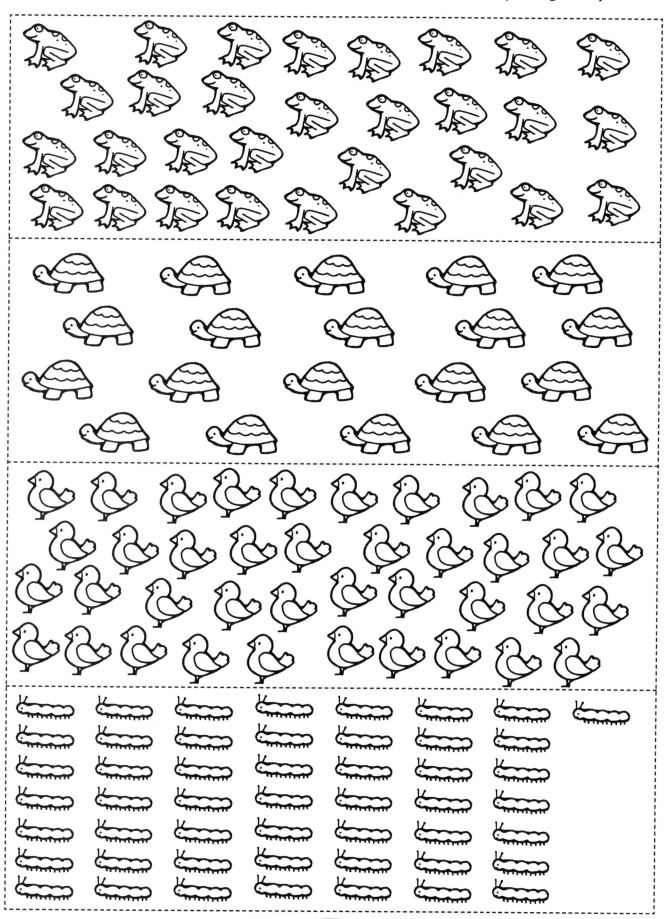

Ideas for Exploring Groups of 2

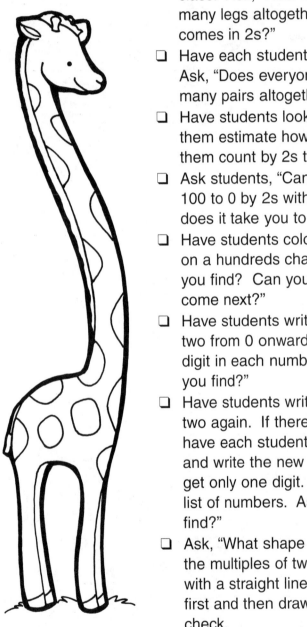

- Ask some students to stand at the front of the class. Ask, "How many pairs of legs? How many legs altogether? What else about bodies comes in 2s?"
- Have each student grab a partner for a dance. Ask, "Does everyone have a match? How many pairs altogether?"
- Have students look at a pile of counters. Have them estimate how many altogether. Have them count by 2s to check.
- Ask students, "Can you count backward from 100 to 0 by 2s without stopping? How long does it take you to do this?"
- Have students color in all the multiples of two on a hundreds chart. Ask, "What pattern can you find? Can you predict which numbers will come next?"
- Have students write down all the multiples of two from 0 onward. Have them look at the last digit in each number. Ask, "What pattern can you find?"
- Have students write down all the multiples of two again. If there are two digits (e.g., 26), have each student add the digits (e.g., 2 + 6) and write the new number (e.g., 8) until they get only one digit. Have them look at the new list of numbers. Ask, "What pattern can you find?"
- Ask, "What shape do you make if you join all the multiples of two in order on a clockface with a straight line?" Have students predict first and then draw a clockface on paper to check.
- Copy page 71 and cut up into 10 cards. Discuss each card together and have students find a way to check by modelling, drawing, or using mental skills. Have them make up more cards for another team to solve.

Ideas for Exploring Groups of 10

- Have students trace around both hands, cut them out, and glue onto a class display about 10. Ask, "How many fingers altogether?" Have students estimate first and then check.
- Have students collect 10-letter words. Ask, "How many can you discover in one minute?"
- Have students find and record things as long as 10 craft sticks. Ask, "How many sticks would you need end-to-end if you placed your things in a long line?"
- Have students collect dimes for a charity group. Ask, "How many coins do you have at the end of each week? How much money have you collected altogether?"
- Ask, "How many different decagons (10-sided figures) can you draw in three minutes?" Have students estimate how many sides altogether, then find a way to check.
- Have students do 10 jumping jacks. Ask, "How many jumping jacks is that if everyone in the class does them at the same time?"
- Have students grab a handful of beansticks. Have them estimate how many beans altogether and then count by 10s to check.
- Have a class challenge. Have each student race another contestant to count by 10s from 0 to 100 and then back to 0 again. Ask, "What's the fastest time you can do this in?"
- Have students count silently by 10s as a leader claps his or her hands. Have students call out the multiple of 10 the leader is on when the leader chooses to stop clapping.

Ideas for Exploring Groups of 5

- Have students make monsters by icing biscuits and decorating with five round candies for eyes and five jelly beans for hair.
- Have students decorate a paper plate. Have them pretend it is a pizza. Have them cut it into five equal pieces. Have students count how many pizza pieces are altogether in their group.
- Have students trace around their hands, cut out, and glue on a long class strip of paper. Have them count by fives to see how many fingers there are altogether.
- Have students stick dots on paper cards in domino patterns. Have them use these to count by fives.
- Have students make a class news book with drawings and stories about groups of five. Have them add a page a day and then find out how many things altogether.
- Have students find out how far they can go on the calculator by pressing 5 + + then = for as many times as they like.
- Have students make small chocolate desserts and sell them at recess for 5 cents each. Ask, "How many did you sell? How much money did you make altogether?"
- Have each student bring a five-layered sandwich for lunch. Ask, "How many slices of bread in the whole class?"
- Have students draw lots of different pentagons. Have them estimate how many sides altogether and then count and check.

Ideas for Exploring Groups — Exploring Multiplication

At the barbeque, there are 8 plates with 5 hot dogs on each plate. How many hot dogs?

I am 4 x 5 plus 2 x 10. What number am I?

There are 7 rabbits. Each one has 2 bunnies. How many bunnies is that?

There are 4 fat cats. Each one has 10 large whiskers. How many whiskers altogether?

I have 6 dogs. Each dog has 10 fleas. How many fleas altogether?

There are 2 haunted houses. Each house has 8 ghosts. How many ghosts is that?

I am 2 x 9 plus 7 x 2. What number am I?

Mom has 9 flower pots. Each pot has 2 flowers in it. How many flowers altogether?

There are 2 boats. Each boat has 6 sailors. How many sailors altogether?

Grandad eats 3 oranges a day. How many oranges will he eat in 10 days?

Exploring Multiplication

Throw and Draw

Skills

- Draw multiplication problems as equal groups or rows. (M1)
- Record using written number sentences. (M5)
- Estimate answers to story problems. (M6)

Grouping

- Work in groups of two to four students.

Materials for Each Group

- Throw and Draw cards for each player (page 73 photocopied and cut up into six cards)
- 0–9 spinner (page 11)
- paper, pencils, and glue for recording

Directions

- Have students select a Throw and Draw number for this round—groups of two (draw two legs on each person), five (draw five fingers on each hand) or ten (draw ten strands of hair on each head).
- Each person takes a matching Throw and Draw card. In turn, use the spinner to find out how many objects to draw. If you select 0, you can call it 0 or 10.

    e.g., Draw 4 groups of 2 legs.

    e.g., Draw 4 groups of 10 strands of hair.

- Have students glue the card on their papers and write the matching number sentence.

    e.g., 4 x 5 = 20

Variation

- Have students draw in legs/fingers/hair on a card. Have each student show it to a partner for about two seconds and then hide it. Ask, "Can your partner predict the matching number sentence?"

Throw and Draw **Exploring Multiplication**

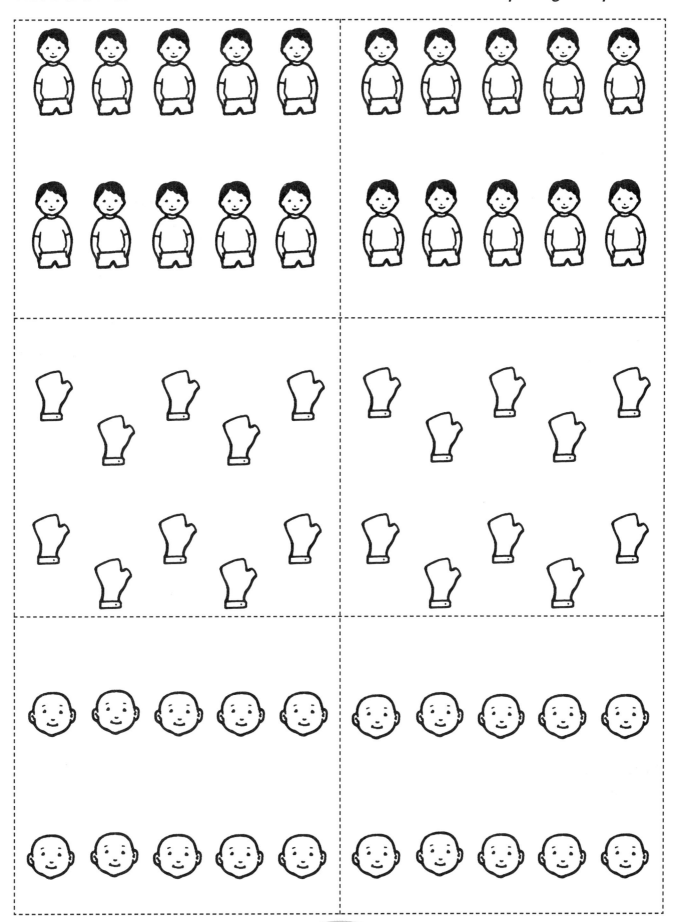

73

Exploring Multiplication

Fat Cat Facts

Skill
- Recall facts for x2, x5, x10 and 2x (doubles). (M7-10)
- Estimate answers to story problems. (M6)
- Record using written number sentences. (M5)

Grouping
- Work in four groups.

Materials
- four copies of the Fat Cat cards (page 75 copied preferably on cardstock and cut up into 10 cards, with each group having a different color)
- pencils

Directions
- Select x2, x5, x10, or 2x facts for each group to record and practice (e.g., x5).
- Each group then allocates the specific facts for each student to record in the middle of each Fat Cat.

e.g.,

- Have students discuss each fact together and how many it represents. You may like students to record the matching answer on the back of each card *(optional)*.
- Have students shuffle all the cards together.
- A leader holds up each card in turn and asks someone to say the complete number fact.

e.g.,  Six groups of five make thirty.

- Repeat until all cards have been used.

Variations
- Group Challenge: Ask, "Who can correctly name all the Fat Cat facts in the shortest time?"
- Have students take personal sets of Fat Cat cards to practice at home.
- Have students make Fat Cat cards for x1, x0, x3, or x4.

Fat Cat Facts Exploring Multiplication

Exploring Multiplication

Crocodile Facts

Skill
- Recall facts for x2, x5, x10. (M7,8,9)

Grouping
- Work as individuals, small groups, or the whole class.

Materials
- a crocodile for each person (page 77 cut into four parts)
- Number Fact strips (page 78 cut up as indicated)
- scissors
- glue

Directions
- Have students decorate their crocodiles. Laminate *(optional)*. Have them cut along the four dashed lines in the center of the crocodile's body.
- Have students take two matching number fact strips.

  (e.g., x2 and x2 answers)
- Have them thread these through the slots in the crocodile's body.
- Have them glue the ends of each strip together to form one continuous line.
- Have students pull each strip through the crocodile's body to find a matching number.

Variations
- Make more copies of the crocodile. Have students write number fact strips for x0, x1, x3 or x4.
- Have students write their own number fact strips for addition or subtraction.

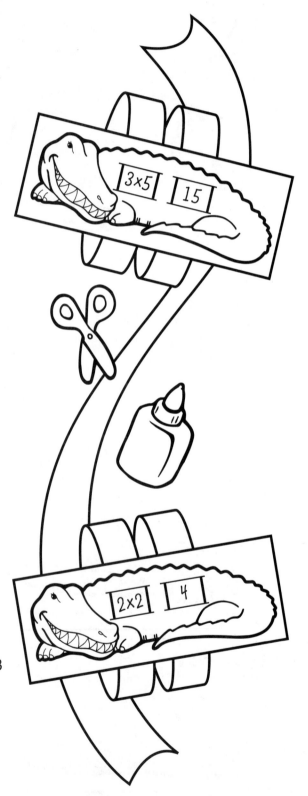

Crocodile Facts   **Exploring Multiplication**

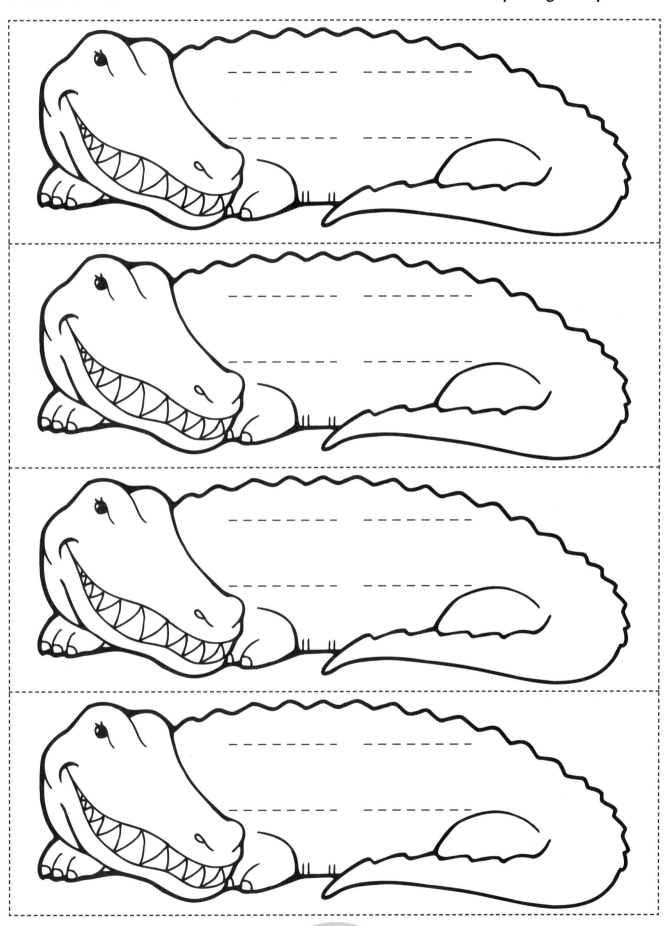

Crocodile Facts — Exploring Multiplication

x 2	x 2 Answers	x 5	x 5 Answers	x 10	x 10 Answers
8 x 2	18	2 x 5	30	10 x 10	0
1 x 2	4	6 x 5	20	0 x 10	30
5 x 2	12	9 x 5	40	8 x 10	70
9 x 2	6	0 x 5	15	1 x 10	60
0 x 2	2	10 x 5	25	9 x 10	10
4 x 2	0	1 x 5	5	3 x 10	40
2 x 2	8	3 x 5	45	5 x 10	80
7 x 2	10	5 x 5	35	7 x 10	20
10 x 2	14	7 x 5	10	2 x 10	50
3 x 2	20	4 x 5	0	6 x 10	90
6 x 2	16	8 x 5	50	4 x 10	100

Exploring Multiplication

Target Challenge

Skill
- ❏ Recall facts for x2, x5, x10, 2x (doubles). (M7,8,9,10)

Grouping
- ❏ Work individually or in pairs.

Materials for Each Group
- ❏ Target Challenge worksheet (page 80 cut into three pieces)
- ❏ pencil
- ❏ one-to-three-minute timer
- ❏ Multiplication Reference Charts (page 83)

Directions
- ❏ Have students write x2, x5, or x10 in the center of each target.
- ❏ Have students write random numbers from 0–10 in the spaces around the center.
- ❏ Have students fill in the outer layer of each target by multiplying the two numbers together each time.
- ❏ Have students keep a record of their progress by filling in the time in minutes or seconds at the bottom of each worksheet.
- ❏ Have students ask a friend to check their answers or use the reference charts.

Variation
- ❏ Have students try targets for x0, x1, x3, or x4. Have students check using the Reference Charts (page 84).

Exploring Multiplication

Beat the Clock

Skill
- Recall facts for x2, x5, x10, 2x (doubles). (M7,8,9,10)

Grouping
- Work individually or in pairs.

Materials for Each Group
- Beat the Clock worksheet (page 82)
- pencil
- one-to-three-minute timer
- Multiplication Reference Charts (page 83)

Directions
- Have students write random numbers from 0–10 in the spaces along the top of each grid.

  e.g.,
x	0	10	3	5	4	8	6	2	7	9

- Have students write 2, 5 or 10 in the spaces on the left of each grid.

  e.g.,
x	
2	

- Have students try to fill in each grid in less than one minute. Have students keep a record of their progress by filling in the number of seconds at the bottom of each grid.
- Have students ask a friend to check their answers or use the reference charts.
- Ask, "Are you faster at answering some facts more than others? What is your fastest time?"

Variations
- Have students try grids for x0, x1, x3 or x4. Check using the Reference Charts (page 84).
- Have students write + in place of the x sign and practice their addition facts.
- Have students write – in place of the x sign and practice their subtraction facts. Have students make sure that each number in the top spaces is larger than the number they are taking away on the left.

Beat the Clock　　　　　　　　　　　　　　　**Exploring Multiplication**

Beat the Clock

Try to fill in each grid in less than 60 seconds.

Number of seconds ....

Number of seconds ....

Number of seconds ....

Number of seconds ....

Number of seconds ....

Number of seconds ....

Beat the Clock   Exploring Multiplication

Groups of 2

0 x 2	0
1 x 2	2
2 x 2	4
3 x 2	6
4 x 2	8
5 x 2	10
6 x 2	12
7 x 2	14
8 x 2	16
9 x 2	18
10 x 2	20

Groups of 10

0 x 10	0
1 x 10	10
2 x 10	20
3 x 10	30
4 x 10	40
5 x 10	50
6 x 10	60
7 x 10	70
8 x 10	80
9 x 10	90
10 x 10	100

Groups of 5

0 x 5	0
1 x 5	5
2 x 5	10
3 x 5	15
4 x 5	20
5 x 5	25
6 x 5	30
7 x 5	35
8 x 5	40
9 x 5	45
10 x 5	50

Doubles

2 x 0	0
2 x 1	2
2 x 2	4
2 x 3	6
2 x 4	8
2 x 5	10
2 x 6	12
2 x 7	14
2 x 8	16
2 x 9	18
2 x 10	20

©Teacher Created Resources, Inc.   83   #3528 Math in Action

Beat the Clock — **Exploring Multiplication**

Groups of 0

0 x 0	0
1 x 0	0
2 x 0	0
3 x 0	0
4 x 0	0
5 x 0	0
6 x 0	0
7 x 0	0
8 x 0	0
9 x 0	0
10 x 0	0

Groups of 1

0 x 1	0
1 x 1	1
2 x 1	2
3 x 1	3
4 x 1	4
5 x 1	5
6 x 1	6
7 x 1	7
8 x 1	8
9 x 1	9
10 x 1	10

Groups of 3

0 x 3	0
1 x 3	3
2 x 3	6
3 x 3	9
4 x 3	12
5 x 3	15
6 x 3	18
7 x 3	21
8 x 3	24
9 x 3	27
10 x 3	30

Groups of 4

0 x 4	0
1 x 4	4
2 x 4	8
3 x 4	12
4 x 4	16
5 x 4	20
6 x 4	24
7 x 4	28
8 x 4	32
9 x 4	36
10 x 4	40

Exploring Multiplication

Multiplication Check-Up

1. Write the matching number sentence.

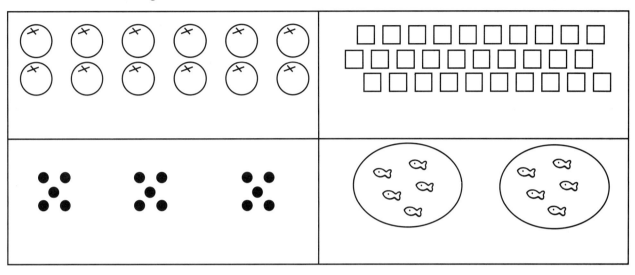

2. Draw balls to show 10 groups of 2. How many balls?

3. Draw sticks to show 4 rows of 5. How many sticks?

4. Write the number sentence to match each story.

　　a. 4 cats with 10 whiskers each _____

　　b. 7 dogs with 5 fleas each _____

　　c. 2 boxes with 8 bananas in each box _____

　　d. 6 pizzas with 2 olives on each one _____

Exploring Multiplication

Multiplication Check-Up

5. Draw and count: Write the number sentence.

a. Five pentagons.
   How many sides?   _____

b. Three bunches of 2 balloons.
   How many balloons?   _____

c. Three faces with 10 freckles on each face.
   How many freckles?   _____

d. Two butterflies with 7 stripes on each wing.
   How many stripes?   _____

6. Write your own story about 3 x 10.

_____

7. Finish these target challenges.

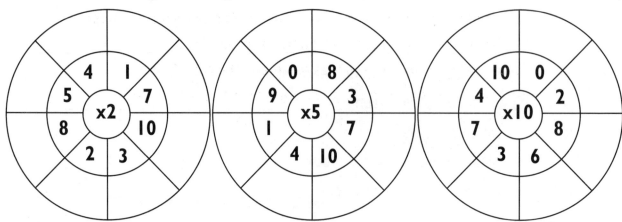

Exploring Division

In this unit, your students will do the following:

- ❑ Recognize shares or groups as equal, fair, unequal, or unfair.
- ❑ Divide a set of objects by sharing or grouping.
- ❑ Understand that sometimes there are objects left over.
- ❑ Use informal language to describe actions.
- ❑ Create and solve division story problems.
- ❑ Estimate answers to story problems.

Exploring Division

What a Lot of Nuts

Skills
- Recognize shares or groups as equal, fair, unequal, or unfair. (D1)
- Divide a set of objects by sharing or grouping. (D2,3)
- Understand that sometimes there are objects left over. (D4)
- Use informal language to describe actions. (D5)
- Create and solve division story problems. (D6)

Grouping
- Work in groups of up to five students.

Materials for Each Group
- ten squirrels (page 89—five copies colored, laminated, and cut out)
- 100 whole nuts (e.g., almonds, peanuts) or dried beans or counters
- What a Lot of Nuts cards (page 90 cut into 10 cards)
- a set of 0–9 digit cards (page 9)
- paper and pencils for recording *(optional)*

Directions
- Have students work as a team. Have them grab a small handful of nuts each and place these together in the center of the group. They do not need to count them. Have students decide whether they will group or share the nuts.
- Have students turn over a digit card (e.g., 6). If grouping, this card tells them how many nuts (e.g., 6) they will give to each squirrel. If sharing, this card tells them how many squirrels they will use for this round (e.g., 6).
- Have students cooperate to model their stories with the nuts. If they run out of squirrels, they may like to keep going anyway!
- Ask, "How many squirrels did you need altogether? How many nuts did each squirrel get? Were there any nuts left over? What will you do with these?"
- Have students find a way to record their stories on paper. Have them draw a picture to match.
- Ask, "How many stories like this can your team solve in five minutes?"

Variations
- Have students use the What a Lot of Nuts cards. Have students estimate the answer first then check by modelling.
- Have students make up their own What a Lot of Nuts cards for another team to solve.

What a Lot of Nuts **Exploring Division**

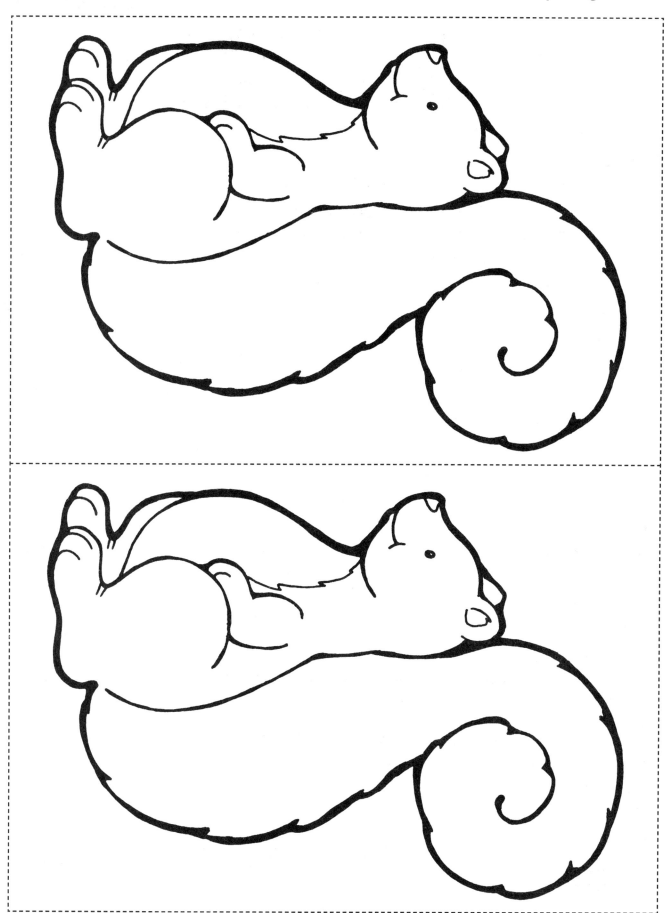

What a Lot of Nuts **Exploring Division**

What a Lot of Nuts   *Exploring Division*

There are 30 nuts to share between 7 squirrels. How many nuts each? How many left over?

Can 5 squirrels take 7 nuts each if there are 36 nuts?

How many squirrels can store exactly 10 nuts if there are 45 nuts altogether? Will there be any nuts left over?

Three squirrels collected nuts in a big pile. They now have 28. How many nuts will they each get? Will any be left over?

Eight squirrels found a pile of 32 nuts. How many will they each get if it is a fair share?

There are 52 nuts. How many squirrels can get 5 nuts each?

There are 47 nuts. How many squirrels can get 6 nuts each?

Some squirrels find 20 nuts and share them fairly. There are no left overs. How many squirrels might there be?

There are 9 squirrels. There are 27 nuts. How many can they each get if it is a fair share?

Four squirrels share some nuts. They each get 7 nuts and there are 2 nuts left over. How many nuts did they have to start with?

Exploring Division

More Ideas for Grouping or Sharing

- ❏ Use up to 100 dragons (pages 7 and 8). Have 1–10 students share the dragons among themselves. Ask, "How many will each of you get?" Have students predict first and then check. (D2,7)

- ❏ Use the What's My Fact? cards (page 17). Have students find how many groups of between 1–10 objects they can circle on each card. Have them predict first and then check. (D3,7)

- ❏ Use the Find My Story cards (page 67). Have students find how many groups of between 1–10 objects they can circle on each strip. Have them predict first and then check. (D3,7)

- ❏ Use the Orchards cards (page 61) and counters (page 60). Have students think of a number between 10 and 100. Have them take this many counters. Have them decide whether they are sharing or grouping and then think of a number between 1 and 10. If grouping, have students find how many trees they will need by the end of their grouping. If sharing, have them find how many pieces of fruit there will be on each tree. Have them predict first and then check. (D2,3,7)

Build It

- ❏ Take a box of building blocks. Have students share the blocks equally between everyone in their group. Ask, "How many do you each get?" On a signal, have students race to build the tallest tower that can stand freely without toppling over.

Bunch Up

- ❏ Work as a whole class outside or in a large hall. On a signal, have students run around as fast as they can. When they hear the whistle blow, have them stop and listen. The leader calls out "Bunch up! Groups of __ (1–10)." Have them race to make groups like that. Any students left over sit out. Continue until there are fewer than five students left.

What a Lot of Freckles

- ❏ Use the faces from page 92. Decide whether students are grouping or sharing. Say, "Now think of a number between 10 and 100." If grouping, have students decide how many freckles (between 1 and 10) they will put on each face. Have them estimate how many faces they will need altogether and whether they will have any extras. Have students check by grouping. If sharing, have students decide how many faces (between 1 and 10) will share freckles. Have them estimate first how many freckles they will end up with on each face and whether there will be any freckles left over. Have students check by sharing.

What a Lot of Spots

- ❏ Use the dogs from page 92. Have students play with the same rules as What a Lot of Freckles. Have students draw spots.

What a Lot of Stripes

- ❏ Use the zebras from page 92. Have students play with the same rules as What a Lot of Freckles. Have students draw stripes.

More Ideas for Grouping or Sharing   **Exploring Division**

More Ideas for Grouping or Sharing   **Exploring Division**

Exploring Division

Division Check-Up

1. Draw in extra freckles so that each child has the same number. How many freckles altogether? _____

2. Share 20 fish equally among 4 aquariums. How many fish in each tank?

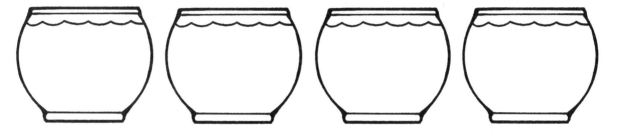

3. How many groups of 7 stars can you make? Are there any stars left over?

4. Draw a picture to show how you would share 15 people equally between 3 cars. How many in each car? Will there be any extras?

Skills Record Sheet

NAME

EXPLORING NUMBERS 0–100 OPERATIONS

Adding	A1	Models facts to 10 with objects or drawings
	A2	Models facts to 20 with objects or drawings
	A3	Creates and solves story problems
	A4	Records activities using symbol/digit cards
	A5	Records activities using written number sentences
	A6	Estimates answers to problems
	A7	Uses a number line to solve addition problems
	A8	Recalls addition facts to 10 using a range of strategies
	A9	Recalls addition facts to 20 using a range of strategies
	A10	Adds three or more digits up to 20
	A11	Models addition to 99 using objects
	A12	Records addition to 99 (without trading)
	A13	Records addition to 99 (with trading)
Subtracting	S1	Models facts to 10 with objects or drawings
	S2	Models facts to 20 with objects or drawings
	S3	Creates and solves story problems
	S4	Records activities using symbol/digit cards
	S5	Records activities using written number sentences
	S6	Estimates answers to problems
	S7	Uses a number line to solve subtraction problems
	S8	Recalls subtraction facts to 10 using a range of strategies
	S9	Recalls subtraction facts to 20 using a range of strategies
	S10	Subtracts three or more digits from up to 20
	S11	Models subtraction to 99 using objects
	S12	Records subtraction to 99 (without trading)
	S13	Records subtraction to 99 (with trading)
Multiplying	M1	Models multiplication as equal groups or rows
	M2	Uses language "groups of," "rows of," and "makes"
	M3	Creates and solves multiplication story problems
	M4	Records multiplication activities using symbol/digit cards
	M5	Records multiplication activities using written number sentences
	M6	Estimates answers to multiplication problems
	M7	Recalls x2 facts
	M8	Recalls x10 facts
	M9	Recalls x5 facts
	M10	Recalls 2x (doubles) facts
	M11	Recalls x1 facts
	M12	Recalls x0 facts
	M13	Recalls x3 facts
	M14	Recalls x4 facts
Dividing	D1	Recognizes shares as "equal"/"fair" or "unequal"/"unfair"
	D2	Divides a set of objects by sharing
	D3	Divides a set of objects by grouping
	D4	Understands that sometimes there are objects left over
	D5	Uses informal language to describe actions
	D6	Creates and solves division story problems
	D7	Estimates answers to simple division story problems

#3528 Math in Action ©Teacher Created Resources, Inc.

Sample Weekly Program

STRAND Number **SUBSTRAND** Addition/Subtraction
GRADE 2 **TERM** 3 **WEEK** 3

LANGUAGE
- and, makes, plus, take away, leaves, equals
- the difference is . . ., altogether make . . ., the total is . . .
- . . . and . . . more makes . . . .
- . . . remove . . . makes . . . .

OUTCOMES
- Group A • Model facts to 10 with objects/drawings (A1, S1)
- Group B • Model facts to 20 with objects/drawings (A2, S2)
- Group C • Add/subtract 3 digits to 20 (A10, S10)
- Groups A,B,C • Record in number sentences (A5, S5)

RESOURCES
Pages 13, 21, 23, 25, 29, 30, 33, 35, 39, 41, 43
overhead transparency (page 17)
transparency pens
two sets of digit cards (page 9) for each student
sets of 20 counters (e.g., blocks, bottle tops)
+ and – symbol cards (page 10)
spinners (page 11)
paper
pencils
scissors
glue

MONDAY	TUESDAY	WEDNESDAY	THURSDAY	FRIDAY
• Whole class challenge: Addition Facts	• Whole class challenge: +/– Facts • Review using Number Line for +/– problems	• Whole class challenge: Subtraction Facts	• Whole class demo: Frog Facts (revision)	• Whole class challenge: What's My Fact? (overhead transparencies)
• **Group A:** Review Tortoise +/– Facts to 10	• **A:** Number Line activities to 10	• **A:** Snake It Away (to 10)	• **A:** Frog Facts pair challenges	• **A:** Tortoise +/– Facts to 10 with a partner
• **Group B:** Review Tortoise +/– Facts to 20	• **B:** Number Line activities to 20	• **B:** Snake It Away (to 20)	• **B:** Grid Challenge, Number Detective	• **B:** Tortoise +/– Facts to 20 with a partner
• **Group C:** Tortoise Snap, Tortoise Challenges Worksheets	• **C:** Number Line activities +/– three digits to 20	• **C:** Caterpillars (– from 20)	• **C:** Grid Challenge, Number Detective	• **C:** Join them up, Life Rafts Puzzle
• Whole class challenge: Addition Facts	• Whole class challenge: +/– Facts	• Whole class challenge: Subtraction Facts	• Whole class challenge: Frog Facts	• Whole class challenge: What's my fact? (overhead transparencies)

Blank Weekly Program

STRAND 　　　　**SUBSTRAND**

GRADE　　　　　**TERM**　　**WEEK**

　　　　　　　　　　LANGUAGE

OUTCOMES

RESOURCES

MONDAY	TUESDAY	WEDNESDAY	THURSDAY	FRIDAY